This is a printed compilation for people that enjoy using and working with printed manuals. The information in this compilation is available for free in PDF format directly from Raspberry Pi. This manual is printed in accordance with their CC BY-ND license. This is a third party printing of their documentation by DienstNet LLC. As an extension, all parts of this compilation not covered by the Raspberry Pi license are also provided under the same CC-BY-ND copyright by DienstNet LLC 2023.

Compilation Contents
Raspberry Pi Pico Pinout
https://datasheets.raspberrypi.com/pico/Pico-R3-A4-Pinout.pdf

Raspberry Pi Pico Product Brief
https://datasheets.raspberrypi.com/pico/pico-product-brief.pdf

Raspberry Pi Pico Datasheet
https://datasheets.raspberrypi.com/pico/pico-datasheet.pdf

ISBN 978-1-365-29210-1

This page was intentionally left blank.

Raspberry Pi Pico Pinout

Raspberry Pi

This page was intentionally left blank.

Raspberry Pi Pico

Published July 2022

Raspberry Pi Ltd

Overview

Raspberry Pi Pico is the debut microcontroller-class board from Raspberry Pi. Built around our RP2040 silicon platform, Pico brings our signature values of high performance, low cost, and ease of use to the microcontroller space.

With a large on-chip memory, symmetric dual-core processor complex, deterministic bus fabric, and rich peripheral set augmented with our unique Programmable I/O (PIO) subsystem, RP2040 provides professional users with unrivalled power and flexibility. With detailed documentation, a polished MicroPython port, and a UF2 bootloader in ROM, it has the lowest possible barrier to entry for beginner and hobbyist users.

RP2040 is manufactured on a modern 40nm process node, delivering high performance, low dynamic power consumption, and low leakage, with a variety of low-power modes to support extended-duration operation on battery power.

Raspberry Pi Pico pairs RP2040 with 2MB of Flash memory, and a power supply chip supporting input voltages from 1.8-5.5V. It provides 26 GPIO pins, three of which can function as analogue inputs, on 0.1"-pitch through-hole pads with castellated edges. Raspberry Pi Pico is available as an individual unit, or in 480-unit reels for automated assembly.

Specification

Form factor:	21 mm × 51 mm
CPU:	Dual-core Arm Cortex-M0+ @ 133MHz
Memory:	264KB on-chip SRAM; 2MB on-board QSPI flash
Interfacing:	26 GPIO pins, including 3 analogue inputs
Peripherals:	• 2 × UART • 2 × SPI controllers • 2 × I2C controllers • 16 × PWM channels • 1 × USB 1.1 controller and PHY, with host and device support • 8 × PIO state machines
Input power:	1.8–5.5V DC
Operating temperature:	-20°C to +85°C
Production lifetime:	Raspberry Pi Pico will remain in production until at least January 2028
Compliance:	For a full list of local and regional product approvals, please visit pip.raspberrypi.com

Physical Specification

Note: all dimensions in mm

WARNINGS

- Any external power supply used with Raspberry Pi Pico shall comply with relevant regulations and standards applicable in the country of intended use.
- This product should be operated in a well-ventilated environment, and if used inside a case, the case should not be covered.
- Whilst in use, this product should be placed on a stable, flat, non-conductive surface, and should not be contacted by conductive items.
- The connection of incompatible devices to Raspberry Pi Pico may affect compliance, result in damage to the unit, and invalidate the warranty.
- All accessories used with this product should comply with relevant standards for the country of use and be marked accordingly to ensure that safety and performance requirements are met.
- The cables and connectors of all peripherals used with this product must have adequate insulation so that relevant safety requirements are met.

SAFETY INSTRUCTIONS

To avoid malfunction or damage to this product, please observe the following:

- Do not expose to water or moisture, or place on a conductive surface whilst in operation.
- Do not expose to heat from any source; Raspberry Pi Pico is designed for reliable operation at normal ambient temperatures.
- Take care whilst handling to avoid mechanical or electrical damage to the printed circuit board and connectors.
- Whilst it is powered, avoid handling the printed circuit board, or only handle it by the corners to minimise the risk of electrostatic discharge damage.

Raspberry Pi is a trademark of Raspberry Pi Ltd

Raspberry Pi Pico Datasheet

An RP2040-based microcontroller board

Colophon

Copyright © 2020-2022 Raspberry Pi Ltd (formerly Raspberry Pi (Trading) Ltd.)

The documentation of the RP2040 microcontroller is licensed under a Creative Commons Attribution-NoDerivatives 4.0 International (CC BY-ND).

build-date: 2022-06-29
build-version: f9f152e-clean

About the SDK

Throughout the text "the SDK" refers to our Raspberry Pi Pico SDK. More details about the SDK can be found in the Raspberry Pi Pico C/C++ SDK book. Source code included in the documentation is Copyright © 2020-2022 Raspberry Pi Ltd (formerly Raspberry Pi (Trading) Ltd.) and licensed under the 3-Clause BSD license.

Legal disclaimer notice

TECHNICAL AND RELIABILITY DATA FOR RASPBERRY PI PRODUCTS (INCLUDING DATASHEETS) AS MODIFIED FROM TIME TO TIME ("RESOURCES") ARE PROVIDED BY RASPBERRY PI LTD ("RPL") "AS IS" AND ANY EXPRESS OR IMPLIED WARRANTIES, INCLUDING, BUT NOT LIMITED TO, THE IMPLIED WARRANTIES OF MERCHANTABILITY AND FITNESS FOR A PARTICULAR PURPOSE ARE DISCLAIMED. TO THE MAXIMUM EXTENT PERMITTED BY APPLICABLE LAW IN NO EVENT SHALL RPL BE LIABLE FOR ANY DIRECT, INDIRECT, INCIDENTAL, SPECIAL, EXEMPLARY, OR CONSEQUENTIAL DAMAGES (INCLUDING, BUT NOT LIMITED TO, PROCUREMENT OF SUBSTITUTE GOODS OR SERVICES; LOSS OF USE, DATA, OR PROFITS; OR BUSINESS INTERRUPTION) HOWEVER CAUSED AND ON ANY THEORY OF LIABILITY, WHETHER IN CONTRACT, STRICT LIABILITY, OR TORT (INCLUDING NEGLIGENCE OR OTHERWISE) ARISING IN ANY WAY OUT OF THE USE OF THE RESOURCES, EVEN IF ADVISED OF THE POSSIBILITY OF SUCH DAMAGE.

RPL reserves the right to make any enhancements, improvements, corrections or any other modifications to the RESOURCES or any products described in them at any time and without further notice.

The RESOURCES are intended for skilled users with suitable levels of design knowledge. Users are solely responsible for their selection and use of the RESOURCES and any application of the products described in them. User agrees to indemnify and hold RPL harmless against all liabilities, costs, damages or other losses arising out of their use of the RESOURCES.

RPL grants users permission to use the RESOURCES solely in conjunction with the Raspberry Pi products. All other use of the RESOURCES is prohibited. No licence is granted to any other RPL or other third party intellectual property right.

HIGH RISK ACTIVITIES. Raspberry Pi products are not designed, manufactured or intended for use in hazardous environments requiring fail safe performance, such as in the operation of nuclear facilities, aircraft navigation or communication systems, air traffic control, weapons systems or safety-critical applications (including life support systems and other medical devices), in which the failure of the products could lead directly to death, personal injury or severe physical or environmental damage ("High Risk Activities"). RPL specifically disclaims any express or implied warranty of fitness for High Risk Activities and accepts no liability for use or inclusions of Raspberry Pi products in High Risk Activities.

Raspberry Pi products are provided subject to RPL's Standard Terms. RPL's provision of the RESOURCES does not expand or otherwise modify RPL's Standard Terms including but not limited to the disclaimers and warranties expressed in them.

Table of contents

- Colophon 1
  - Legal disclaimer notice 1
- 1. About Raspberry Pi Pico 3
  - 1.1. Raspberry Pi Pico design files 5
- 2. Mechanical specification 6
  - 2.1. Raspberry Pi Pico pinout 6
  - 2.2. Surface-mount footprint 8
  - 2.3. Recommended operating conditions 9
- 3. Electrical specification 11
  - 3.1. Power consumption 11
    - 3.1.1. Popcorn 11
    - 3.1.2. BOOTSEL mode 13
    - 3.1.3. DORMANT mode 14
    - 3.1.4. SLEEP mode 15
- 4. Applications information 17
  - 4.1. Programming the flash 17
  - 4.2. General purpose I/O 17
  - 4.3. Using the ADC 17
  - 4.4. Powerchain 18
  - 4.5. Powering Pico 19
  - 4.6. Using a battery charger 20
  - 4.7. USB 21
  - 4.8. Debugging 21
- Appendix A: Availability 22
  - Support 22
  - Ordering code 22
- Appendix B: Pico schematic 23
- Appendix C: Pico component locations 26
- Appendix D: Documentation release history 27

Chapter 1. About Raspberry Pi Pico

Raspberry Pi Pico is a microcontroller board based on the Raspberry Pi RP2040 microcontroller chip.

Figure 1. The Raspberry Pi Pico Rev3 board.

Raspberry Pi Pico has been designed to be a low cost yet flexible development platform for RP2040, with the following key features:

- RP2040 microcontroller with 2MB Flash
- Micro-USB B port for power and data (and for reprogramming the Flash)
- 40 pin 21×51 'DIP' style 1mm thick PCB with 0.1" through-hole pins also with edge castellations
  - Exposes 26 multi-function 3.3V General Purpose I/O (GPIO)
  - 23 GPIO are digital-only and 3 are ADC capable
  - Can be surface mounted as a module
- 3-pin ARM Serial Wire Debug (SWD) port
- Simple yet highly flexible power supply architecture
  - Various options for easily powering the unit from micro-USB, external supplies or batteries
- High quality, low cost, high availability
- Comprehensive SDK, software examples and documentation

For full details of the RP2040 microcontroller please see the RP2040 Datasheet, however the headline features are:

- Dual-core cortex M0+ at up to 133MHz
  - On-chip PLL allows variable core frequency
- 264kB multi-bank high performance SRAM
- External Quad-SPI Flash with eXecute In Place (XIP) and 16kB on-chip cache
- High performance full-crossbar bus fabric

- On-board USB1.1 (device or host)
- 30 multi-function General Purpose IO (4 can be used for ADC)
    - 1.8-3.3V IO Voltage (**NOTE** Pico IO voltage is fixed at 3.3V)
- 12-bit 500ksps Analogue to Digital Converter (ADC)
- Various digital peripherals
    - 2 × UART, 2 × I2C, 2 × SPI, 16 × PWM channels
    - 1 × Timer with 4 alarms, 1 × Real Time Counter
- 2 × Programmable IO (PIO) blocks, 8 state machines total
    - Flexible, user-programmable high-speed IO
    - Can emulate interfaces such as SD Card and VGA

Pico provides minimal (yet flexible) external circuitry to support the RP2040 chip: flash (Winbond W25Q16JV), crystal, power supplies and decoupling, and USB connector. The majority of the RP2040 microcontroller pins are brought to the user IO pins on the left and right edge of the board. Four RP2040 IO are used for internal functions - driving an LED, on-board Switched Mode Power Supply (SMPS) power control and sensing the system voltages.

Pico has been designed to use either soldered 0.1" pin-headers (it is one 0.1" pitch wider than a standard 40-pin DIP package) or can be used as a surface mountable 'module', as the user IO pins are also castellated. There are SMT pads underneath the USB connector and BOOTSEL button, which allow these signals to be accessed if used as a reflow-soldered SMT module.

Figure 2. The pinout of the Raspberry Pi Pico Rev3 board

Pico uses an on-board buck-boost SMPS which is able to generate the required 3.3V (to power RP2040 and external circuitry) from a wide range of input voltages (~1.8 to 5.5V). This allows significant flexibility in powering the unit from various sources such as a single Lithium-Ion cell, or 3 AA cells in series. Battery chargers can also be very easily integrated with the Pico powerchain.

Reprogramming the Pico Flash can be done using USB (simply drag and drop a file onto the Pico which appears as a mass storage device), or the standard Serial Wire Debug (SWD) port can reset the system and load and run code without any button presses. The SWD port can also be used to interactively debug code running on the RP2040.

> **Getting started with Raspberry Pi Pico**
>
> The Getting started with Raspberry Pi Pico book walks through loading programs onto the board, and shows how to install the C/C++ SDK and build the example C programs. See the Raspberry Pi Pico Python SDK book to get started with MicroPython, which is the fastest way to get code running on Pico.

1.1. Raspberry Pi Pico design files

The source design files, including the schematic and PCB layout, are made available openly, with no limitations.

Schematic The full schematic is reproduced in Appendix B. The schematic is also distributed alongside the layout files here.

Layout The full CAD files, including PCB layout, can be found here. Note that Raspberry Pi Pico was designed in Cadence Allegro PCB Editor, and opening in other PCB CAD packages requires an import script or plugin.

STEP 3D A STEP 3D model of Raspberry Pi Pico, for 3D visualisation and fit check of designs which include Raspberry Pi Pico as a module, can be found here.

Fritzing A Fritzing part for use in e.g. breadboard layouts can be found here.

Permission to use, copy, modify, and/or distribute this design for any purpose with or without fee is hereby granted.

THE DESIGN IS PROVIDED "AS IS" AND THE AUTHOR DISCLAIMS ALL WARRANTIES WITH REGARD TO THIS DESIGN INCLUDING ALL IMPLIED WARRANTIES OF MERCHANTABILITY AND FITNESS. IN NO EVENT SHALL THE AUTHOR BE LIABLE FOR ANY SPECIAL, DIRECT, INDIRECT, OR CONSEQUENTIAL DAMAGES OR ANY DAMAGES WHATSOEVER RESULTING FROM LOSS OF USE, DATA OR PROFITS, WHETHER IN AN ACTION OF CONTRACT, NEGLIGENCE OR OTHER TORTIOUS ACTION, ARISING OUT OF OR IN CONNECTION WITH THE USE OR PERFORMANCE OF THIS DESIGN.

> **Other RP2040 design resources**
>
> The Hardware design with RP2040 book walks through designing a minimal RP2040-based 2-layer board (KiCad files here), and another board which uses Raspberry Pi Pico as a module (KiCad files here), explaining the relevant design considerations along the way, at both the schematic and layout level.

Chapter 2. Mechanical specification

The Raspberry Pi Pico is a single sided 51×21mm 1mm thick PCB with a micro-USB port overhanging the top edge and dual castellated/through-hole pins around the remaining edges. Pico is designed to be usable as a surface mount module as well as being in Dual Inline Package (DIP) type format, with the 40 main user pins on a 2.54mm (0.1") pitch grid with 1mm holes and hence compatible with veroboard and breadboard. Pico also has 4× 2.1mm (± 0.05mm) drilled mounting holes to provide for mechanical fixing, see Figure 3.

Figure 3. The dimensions of the Raspberry Pi Pico Rev3 board.

2.1. Raspberry Pi Pico pinout

The Pico pinout has been designed to directly bring out as much of the RP2040 GPIO and internal circuitry function as possible, while also providing a suitable number of ground pins to reduce EMI (Electro Magnetic Interference) and signal crosstalk. This is important in general but especially for RP2040 which is built on a modern 40nm silicon process and hence the digital IO edge rates are very fast.

Figure 4. The pin numbering of the Raspberry Pi Pico Rev3 board.

> **NOTE**
>
> The physical pin numbering is shown in Figure 4, for the pin allocation see Figure 2 or the full Raspberry Pi Pico schematics in Appendix B.

A few RP2040 GPIO pins are used for internal board functions, these are:

GPIO29 IP Used in ADC mode (ADC3) to measure VSYS/3

GPIO25 OP Connected to user LED

GPIO24 IP VBUS sense - high if VBUS is present, else low

GPIO23 OP Controls the on-board SMPS Power Save pin (Section 4.4)

Apart from GPIO and ground pins, there are 7 other pins on the main 40-pin interface:

PIN40 VBUS

PIN39 VSYS

PIN37 3V3_EN

PIN36 3V3

PIN35 ADC_VREF

PIN33 AGND

PIN30 RUN

VBUS is the micro-USB input voltage, connected to micro-USB port pin 1. This is nominally 5V (or 0V if the USB is not connected or not powered).

VSYS is the main system input voltage, which can vary in the allowed range 1.8V to 5.5V, and is used by the on-board

SMPS to generate the 3.3V for the RP2040 and its GPIO.

3V3_EN connects to the on-board SMPS enable pin, and is pulled high (to VSYS) via a 100kΩ resistor. To disable the 3.3V (which also de-powers the RP2040), short this pin low.

3V3 is the main 3.3V supply to RP2040 and its I/O, generated by the on-board SMPS. This pin can be used to power external circuitry (maximum output current will depend on RP2040 load and VSYS voltage, it is recommended to keep the load on this pin less than 300mA).

ADC_VREF is the ADC power supply (and reference) voltage, and is generated on Pico by filtering the 3.3V supply. This pin can be used with an external reference if better ADC performance is required.

AGND is the ground reference for GPIO26-29, there is a separate analog ground plane running under these signals and terminating at this pin. If the ADC is not used or ADC performance is not critical, this pin can be connected to digital ground.

RUN is the RP2040 enable pin, and has an internal (on-chip) pull-up resistor to 3.3V of about ~50kΩ. To reset RP2040, short this pin low.

Finally, there are also 6 Test Points (TP1-TP6) which can be accessed if required, for example if using as a surface mount module. These are:

TP1   Ground (close coupled ground for differential USB signals)

TP2   USB DM

TP3   USB DP

TP4   GPIO23/SMPS PS pin (do not use)

TP5   GPIO25/LED (not recommended to be used)

TP6   BOOTSEL

TP1, TP2 and TP3 can be used to access the USB signals instead of using the micro-USB port. TP6 can be used to drive the system into mass-storage USB programming mode (by shorting it low at power-up). Note that TP4 is not intended to be used externally, and TP5 is not really recommended to be used as it will only swing from 0V to the LED forward voltage (and hence can only really be used as an output with special care).

2.2. Surface-mount footprint

The following footprint (Figure 5) is recommended for systems which will be reflow-soldering Pico units as modules.

Figure 5. The SMT footprint of the Raspberry Pi Pico Rev3 board.

The footprint shows the test point locations and pad sizes as well as the 4 USB connector shell ground pads (A,B,C,D). The USB connector on Pico is a through-hole part, which provides it with mechanical strength. The USB socket pins do not protrude all the way through the board, however solder does pool at these pads during manufacture and can stop the module sitting completely flat. Hence we provide pads on the SMT module footprint to allow this solder to reflow in a controlled manner when Pico goes through reflow again.

For test points that are not used, it is acceptable to void any copper under these (with suitable clearance) on the carrier board.

2.3. Recommended operating conditions

Operating conditions for the Raspberry Pi Pico are largely a function of the operating conditions specified by its components.

Operating Temp Max	85°C (including self-heating)
Operating Temp Min	-20°C
VBUS	5V ± 10%.
VSYS Min	1.8V
VSYS Max	5.5V

Note that VBUS and VSYS current will depend on use-case, some examples are given in the next section.

Recommended maximum ambient temperature of operation is 70°C.

Chapter 3. Electrical specification

3.1. Power consumption

The power consumption tables and graphs show the VBUS current consumption on three typical Raspberry Pi Pico devices, with four different software use cases. These results are not guaranteed maximum values; they are an indication of the current consumption a user can typically expect the device to draw when used in these scenarios.

For more detailed current consumption data, please see the RP2040 Datasheet.

3.1.1. Popcorn

The 'Popcorn' media player demo uses the VGA, SD card, and audio board. This demo uses VGA video, I2S audio and 4-bit SD card access. This has been performed both with power-saving mode (Table 1), and without power-saving mode (Table 2). The power saving (PS) mode of Raspberry Pi Pico's voltage regulator is controlled by GPIO23. The 'Average' data shows the mean current consumption over several seconds of video, with varying colour and intensity. The 'Maximum' data is obtained during periods of white video, when the current required is at its highest.

Table 1. Popcorn running on the VGA board (VGA video, SD card and I2S audio)

Pico board	Average VBUS current @5V (mA)			Maximum VBUS current @5V (mA)		
	Temperature (°C)			Temperature (°C)		
	-25	25	85	-25	25	85
#1	84.8	83.4	87.3	90.9	87.8	93.1
#2	87.5	89.9	89.8	93.4	94.1	94.0
#3	84.4	86.2	86.9	90.6	92.9	91.3
Mean	**85.6**	**86.5**	**88.0**	**91.6**	**91.6**	**92.8**

> **NOTE**
>
> Includes current consumed by VGA board (e.g. I2S DAC) in addition to the Raspberry Pi Pico. For more information on the VGA board see the Hardware design with RP2040 book.

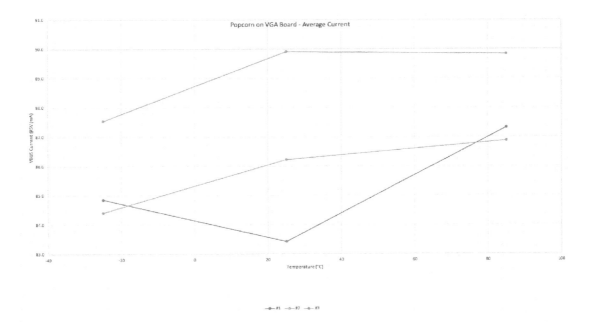

Figure 6. Popcorn running on the VGA board (average current)

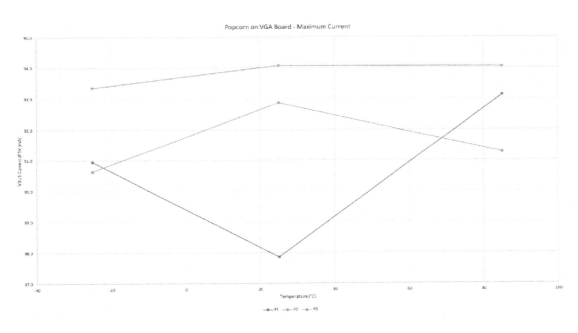

Figure 7. Popcorn running on the VGA board (maximum current)

*Table 2. Popcorn running on the VGA board (VGA video, SD card and I2S audio) with power saving **disabled** (GPIO23 pulled high)*

Pico board	Average VBUS current @5V (mA)			Maximum VBUS current @5V (mA)		
	Temperature (°C)			Temperature (°C)		
	-25	25	85	-25	25	85
#1	90.3	89.3	90.7	94.6	94.5	95.6
#2	91.2	87.1	90.6	95.4	91.0	95.5
#3	90.8	90.2	91.6	95.0	95.1	95.3
Mean	**90.8**	**88.9**	**91.0**	**95.0**	**93.5**	**95.5**

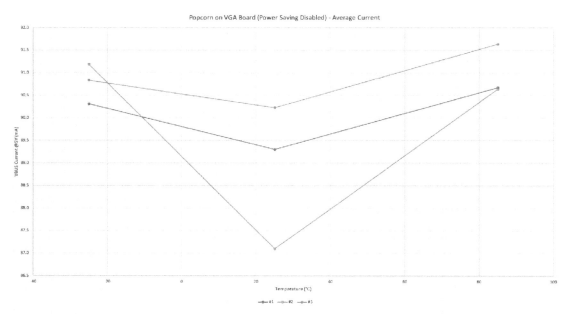

Figure 8. Popcorn running on the VGA board (average current) with power saving **disabled** (`GPIO23` pulled high)

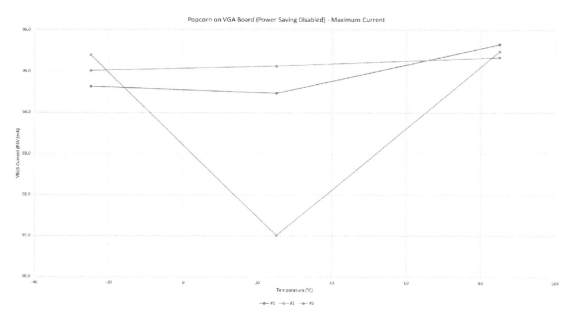

Figure 9. Popcorn running on the VGA board (maximum current) with power saving **disabled** (`GPIO23` pulled high)

3.1.2. BOOTSEL mode

Table 3 describes the BOOTSEL mode of RP2040. These measurements are made both with and without USB activity on the bus, using a Raspberry Pi 4 as a host.

Table 3. BOOTSEL mode

Pico board	USB idle VBUS current @5V (mA)			USB active VBUS current @5V (mA)		
	Temperature (°C)			Temperature (°C)		
	-25	25	85	-25	25	85
#1	9.05	8.50	8.87	10.19	9.66	10.00
#2	9.58	9.12	9.20	10.64	10.35	10.36
#3	9.45	8.63	8.99	10.75	9.81	10.08

| Mean | 9.4 | 8.7 | 9.0 | 10.5 | 9.9 | 10.1 |

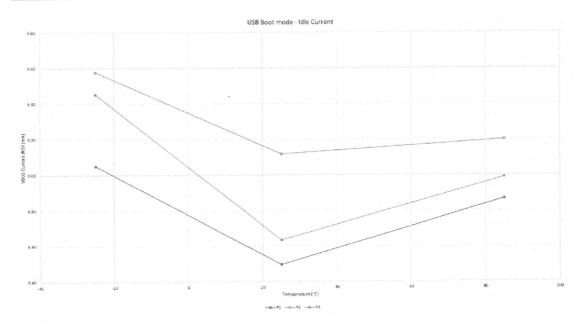

Figure 10. BOOTSEL mode, idle current.

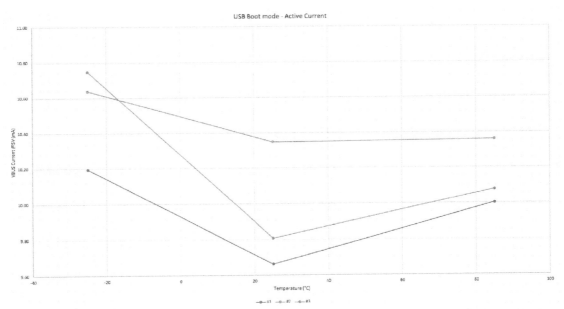

Figure 11. BOOTSEL mode, active current.

3.1.3. DORMANT mode

This use-case employs the `hello_dormant` binary which puts RP2040 into DORMANT mode, a very low-power state.

Table 4. Raspberry Pi Pico running the `hello_dormant` binary. Board is in DORMANT mode

Pico board	VBUS current @5V (mA)		
	Temperature (°C)		
	-25	25	85
#1	1.16	0.78	1.34
#2	1.19	0.92	1.40

3.1. Power consumption

#3	1.29	0.85	1.33
Mean	**1.2**	**0.8**	**1.4**

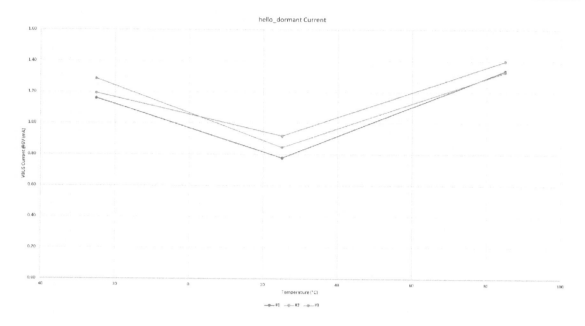

Figure 12. Raspberry Pi Pico running the `hello_dormant` binary. Board is in DORMANT mode.

3.1.4. SLEEP mode

The final example uses the `hello_sleep` binary code, which puts RP2040 into SLEEP mode, a low power state which leaves some clock infrastructure in the chip turned on (which is not the case in the lower-power DORMANT mode).

Table 5. Raspberry Pi Pico running the `hello_sleep` binary. Board is in SLEEP mode.

Pico board	VBUS current @5V (mA)		
	Temperature (°C)		
	-25	25	85
#1	1.35	1.30	1.81
#2	1.53	1.39	1.92
#3	1.40	1.32	1.92
Mean	**1.4**	**1.3**	**1.9**

Figure 13. Raspberry Pi Pico running the `hello_sleep` *binary. Board is in* SLEEP *mode.*

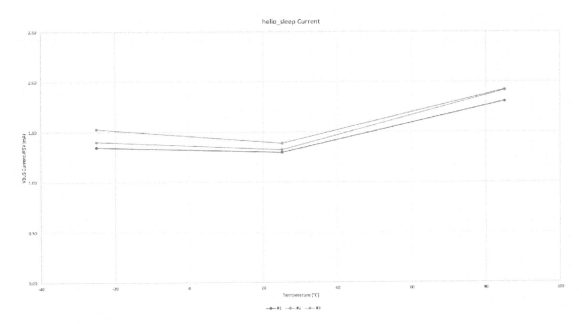

Chapter 4. Applications information

4.1. Programming the flash

The on-board 2MB QSPI Flash can be (re)programmed either using the Serial Wire Debug port or by the special USB Mass Storage Device mode.

The simplest way to reprogram the Pico's Flash is to use the USB mode. To do this, depower the board, then hold the BOOTSEL button down during board power-up (e.g. hold BOOTSEL down while connecting the USB). The Pico will then appear as a USB Mass Storage Device. Dragging a special '.uf2' file onto the disk will write this file to the Flash and restart the Pico.

The USB boot code is stored in ROM on RP2040, so can not be accidentally overwritten.

To get started using the SWD port see the Debugging with SWD section in the Getting started with Raspberry Pi Pico book.

4.2. General purpose I/O

The Raspberry Pi Pico's GPIO is powered from the on-board 3.3V rail and is therefore fixed at 3.3V.

The Pico exposes 26 of the 30 possible RP2040 GPIO pins by routing them straight out to Pico header pins. GPIO0 to GPIO22 are digital only and GPIO 26-28 are able to be used either as digital GPIO or as ADC inputs (software selectable).

One thing to note is that the ADC capable GPIO26-29 have an internal reverse diode to the VDDIO (3V3) rail and so the input voltage must not exceed VDDIO plus about 300mV. Also, if the RP2040 is unpowered, applying a voltage to these GPIO pins will 'leak' through the diode into the VDDIO rail. Normal digital GPIO pins 0-25 (and also the debug pins) do not have this restriction and therefore voltage can safely be applied to these pins when RP2040 is unpowered.

4.3. Using the ADC

The RP2040 ADC does not have an on-board reference and therefore uses its own power supply as a reference. On Pico the ADC_AVDD pin (the ADC supply) is generated from the SMPS 3.3V by using an R-C filter (201Ω into 2.2µF). This is a simple solution but does have the following drawbacks:

1. We are relying on the 3.3V SMPS output accuracy which isn't great
2. We can only do so much filtering and therefore ADC_AVDD will be somewhat noisy
3. The ADC draws current (about 150µA if the temperature sense diode is disabled, but it varies from chip to chip) and therefore there will be an inherent offset of about 150µA*200 = ~30mV. There is a small difference in current draw when the ADC is sampling (about +20µA) so that offset will also vary with sampling as well as operating temperature.

Changing the resistance between the ADC_VREF and 3V3 pin can reduce the offset at the expense of more noise - which may be OK especially if the use case can support averaging over multiple samples.

Driving high the SMPS mode pin (GPIO23), to force the power supply into PWM mode, can greatly reduce the inherent ripple of the SMPS at light load, and therefore the ripple on the ADC supply. This does reduce the power efficiency of the board at light load, so the low-power PFM mode can be re-enabled between infrequent ADC measurements by driving GPIO23 low once more. See Section 4.4.

The ADC offset can be reduced by tying a second channel of the ADC to ground, and using this zero-measurement as an approximation to the offset.

For much improved ADC performance, an external 3.0V shunt reference, such as LM4040, can be connected from the ADC_VREF pin to ground. Note that if doing this the ADC range is limited to 0-3.0V signals (rather than 0-3.3V), and the shunt reference will draw continuous current through the 200Ω filter resistor (3.3V-3.0V)/200 = ~1.5mA.

Note that the 1Ω resistor on Pico (R9) is designed to (maybe) help with shunt references that would otherwise become unstable when directly connected to 2.2µF. It also makes sure there is a little filtering even in the case that 3.3V and ADC_VREF are shorted together (which is a valid thing to do if you don't care about noise and want to reduce the inherent offset).

Finally, R7 is a physically large 1608 metric (0603) package resistor, so can be relatively easily removed if a user wants to isolate ADC_VREF and do their own thing with the ADC voltage, for example powering it from an entirely separate voltage (e.g. 2.5V). Note that the ADC on RP2040 has only been qualified at 3.0/3.3V but should work down to about 2V.

4.4. Powerchain

Raspberry Pi Pico has been designed with a simple yet flexible power supply architecture and can easily be powered from other sources such as batteries or external supplies. Integrating the Pico with external charging circuits is also straightforward. Figure 14 shows the power supply circuitry.

Figure 14. The powerchain of the Raspberry Pi Pico Rev3 board.

VBUS is the 5V input from the micro-USB port, which is fed through a Schottky diode to generate VSYS. The VBUS to VSYS diode (D1) adds flexibility by allowing power ORing of different supplies into VSYS.

VSYS is the main system 'input voltage' and feeds the RT6150 buck-boost SMPS, which generates a fixed 3.3V output for the RP2040 device and its IO (and can be used to power external circuitry). VSYS is R-C filtered and divided by 3 (by R5, R6 and C3 in the Pico schematic) and can be monitored on ADC channel 3. This can be used for example as a crude battery voltage monitor.

The buck-boost SMPS, as its name implies, can seamlessly switch from buck to boost mode, and therefore can maintain an output voltage of 3.3V from a wide range of input voltages, ~1.8V to 5.5V, which allows a lot of flexibility in the choice of power source.

GPIO24 monitors the existence of VBUS, while R10 and R1 act to pull VBUS down to make sure it is 0V if VBUS is not present.

GPIO23 controls the RT6150 PS (Power Save) pin. When PS is low (the default on Pico) the regulator is in Pulse Frequency Modulation mode, which, at light loads, saves considerable power by only turning on the switching MOSFETs occasionally to keep the output capacitor topped up. Setting PS high forces the regulator into Pulse Width Modulation (PWM) mode. PWM mode forces the SMPS to switch continuously, which reduces the output ripple considerably at light loads (which can be good for some use cases) but at the expense of much worse efficiency. Note that under heavy load the switcher will be in PWM mode irrespective of the PS pin state.

The SMPS EN pin is pulled up to VSYS by a 100kΩ resistor and made available on Pico pin 37. Shorting this pin to ground will disable the switcher and put it into a low power state.

> **ℹ NOTE**
>
> The RP2040 has an on-chip linear regulator (LDO) that powers the digital core at 1.1V (nominal) from the 3.3V supply, which is not shown in Figure 14.

4.5. Powering Pico

The simplest way to power Pico is to plug in the micro-USB, which will power VSYS (and therefore the system) from the 5V USB VBUS voltage, via D1 (so VSYS becomes VBUS minus the Schottky diode drop).

If the USB port is the **only** power source, VSYS and VBUS can be safely shorted together to eliminate the Schottky diode drop (which improves efficiency and reduces ripple on VSYS).

If the USB port is **not** going to be used, it is safe to power Pico by connecting VSYS to your preferred power source (in the range ~1.8V to 5.5V).

> **❗ IMPORTANT**
>
> If you are using Raspberry Pi Pico in USB Host mode (e.g. using one of the TinyUSB host examples) then you must power Pico by providing 5V to the VBUS pin.

The simplest way to safely add a second power source to Pico is to feed it into VSYS via another Schottky diode (see Figure 15). This will 'OR' the two voltages, allowing the higher of either the external voltage or VBUS to power VSYS, with the diodes preventing either supply from back-powering the other. For example a single Lithium-Ion cell* (cell voltage ~3.0V to 4.2V) will work well, as will 3×AA series cells (~3.0V to ~4.8V) and any other fixed supply in the range ~2.3V to 5.5V. The downside of this approach is that the second power supply will suffer a diode drop in the same way as VBUS does, and this may not be desirable from an efficiency perspective or if the source is already close to the lower range of input voltage allowed for the RT6150.

Figure 15. Raspberry Pi Pico power ORing using diodes.

An improved way to power from a second source is using a P-channel MOSFET (P-FET) to replace the Schottky diode as shown in Figure 16. Here, the gate of the FET is controlled by VBUS, and will disconnect the secondary source when VBUS is present. The P-FET should be chosen to have low on resistance, and therefore overcomes the efficiency and voltage-drop issues with the diode-only solution.

Note that the V_t (threshold voltage) of the P-FET must be chosen to be well below the minimum external input voltage, to make sure the P-FET is turned on swiftly and with low resistance. When the input VBUS is removed, the P-FET will not start to turn on until VBUS drops below the P-FET's V_t, meanwhile the body diode of the P-FET may start to conduct (depending on whether V_t is smaller than the diode drop). For inputs that have a low minimum input voltage, or if the P-FET gate is expected to change slowly (e.g. if any capacitance is added to VBUS) a secondary Schottky diode across the P-FET (in the same direction as the body diode) is recommended. This will reduce the voltage drop across the P-FET's body diode.

An example of a suitable P-MOSFET for most situations is Diodes DMG2305UX which has a maximum V_t of 0.9V and R_{on} of 100mΩ (at 2.5V V_{gs}).

Figure 16. Raspberry Pi Pico power ORing using P channel MOSFET.

⚠ **CAUTION**

If using Lithium-Ion cells they must have, or be provided with, adequate protection against over-discharge, over-charge, charging outside allowed temperature range, and overcurrent. Bare, unprotected cells are dangerous and can catch fire or explode if over-discharged, over-charged or charged / discharged outside their allowed temperature and/or current range.

4.6. Using a battery charger

Pico can also be used with a battery charger. Although this is a slightly more complex use case it is still straightforward. Figure 17 shows an example of using a 'Power Path' type charger (where the charger seamlessly manages swapping between powering from battery or powering from the input source and charging the battery, as needed).

Figure 17. Using Raspberry Pi Pico with a charger.

In the example we feed VBUS to the input of the charger, and we feed VSYS with the output via the previously mentioned P-FET arrangement. Depending on your use case you may also want to add a Schottky diode across the P-FET as described in the previous section.

4.7. USB

RP2040 has an integrated USB1.1 PHY and controller which can be used in both Device and Host mode. Pico adds the two required 27Ω external resistors and brings this interface to a standard micro-USB port.

The USB port can be used to access the USB bootloader (BOOTSEL mode) stored in the RP2040 boot ROM. It can also be used by user code, to access an external USB device or host.

4.8. Debugging

Raspberry Pi Pico brings the RP2040 Serial Wire Debug (SWD) interface to a 3 pin debug header on the lower edge of the board. To get started using the debug port see the Debugging with SWD section in the Getting started with Raspberry Pi Pico book.

 NOTE

The RP2040 chip has internal pull up resistors on the SWDIO and SWCLK pins, both nominally 60kΩ.

Appendix A: Availability

Raspberry Pi guarantee availability of the Raspberry Pi Pico product until at least January 2028.

Support

For support see the Pico section of the Raspberry Pi website, and post questions on the Raspberry Pi forum.

Ordering code

Table 6. Part Number

Model	Order Code	EAN	Minimal Order Quantity	RRP
Raspberry Pi Pico	SC0915	5056561801445	1+ pcs / Bulk	US$4.00
	SC0916	0617588405587	1+ pcs / Bulk	US$4.00
Raspberry Pi Pico H	SC0917	5056561803180	1+ pcs / Bulk	US$5.00

 NOTE

RRP was correct at time of publication and excludes taxes.

Appendix B: Pico schematic

See Figure 18 on the following page.

Figure 18. The Raspberry Pi Pico Rev3 board schematic.

Appendix C: Pico component locations

Figure 19. The Raspberry Pi Pico Rev3 board component locations.

Appendix D: Documentation release history

Table 7. Documentation release history

Release	Date	Description
1.0	21 Jan 2021	• Initial release
1.1	26 Jan 2021	• Minor corrections • Extra information about using DMA with ADC • Clarified M0+ and SIO CPUID registers • Added more discussion of Timers • Update Windows and macOS build instructions • Renamed books and optimised size of output PDFs
1.2	01 Feb 2021	• Minor corrections • Small improvements to PIO documentation • Added missing TIMER2 and TIMER3 registers to DMA • Explained how to get MicroPython REPL on UART • To accompany the V1.0.1 release of the C SDK
1.3	23 Feb 2021	• Minor corrections • Changed font • Additional documentation on sink/source limits for RP2040 • Major improvements to SWD documentation • Updated MicroPython build instructions • MicroPython UART example code • Updated Thonny instructions • Updated Project Generator instructions • Added a FAQ document • Added errata E7, E8 and E9
1.3.1	05 Mar 2021	• Minor corrections • To accompany the V1.1.0 release of the C SDK • Improved MicroPython UART example • Improved Pinout diagram
1.4	07 Apr 2021	• Minor corrections • Added errata E10 • Note about how to update the C SDK from Github • To accompany the V1.1.2 release of the C SDK

Release	Date	Description
1.4.1	13 Apr 2021	• Minor corrections • Clarified that all source code in the documentation is under the 3-Clause BSD license.
1.5	07 Jun 2021	• Minor updates and corrections • Updated FAQ • Added SDK release history • To accompany the V1.2.0 release of the C SDK
1.6	23 Jun 2021	• Minor updates and corrections • ADC information updated • Added errata E11
1.6.1	30 Sep 2021	• Minor updates and corrections • Information about B2 release • Updated errata for B2 release
1.7	03 Nov 2021	• Minor updates and corrections • Fixed some register access types and descriptions • Added core 1 launch sequence info • Described SDK "panic" handling • Updated picotool documentation • Additional examples added to **Appendix A: App Notes** appendix in the Raspberry Pi Pico C/C++ SDK book • To accompany the V1.3.0 release of the C SDK
1.7.1	04 Nov 2021	• Minor updates and corrections • Better documentation of USB double buffering • Picoprobe branch changes • Updated links to documentation
1.8	17 Jun 2022	• Minor updates and corrections • Updated setup instructions for Windows in Getting started with Raspberry Pi Pico • Additional explanation of SDK configuration • RP2040 now qualified to -40°C, minimum operating temperature changed from -20°C to -40°C • Increased PLL min VCO from 400MHz to 750MHz for improved stability across operating conditions • Added reflow-soldering temperature profile • Added errata E12, E13 and E14 • To accompany the V1.3.1 release of the C SDK

Release	Date	Description
1.9	30 Jun 2022	- Minor updates and corrections - Update to VGA board hardware description for launch of Raspberry Pi Pico W - To accompany the V1.4.0 release of the C SDK

The latest release can be found at https://datasheets.raspberrypi.com/pico/pico-datasheet.pdf.

Raspberry Pi is a trademark of Raspberry Pi Ltd

Raspberry Pi Ltd

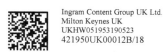
Ingram Content Group UK Ltd.
Milton Keynes UK
UKHW051953190523
421950UK00012B/18